FIRE

DISASTERS

Laura Conlon

The Rourke Corporation, Inc.
Vero Beach, Florida 32964

Edited by Sandra A. Robinson

PHOTO CREDITS
© Tom and Pat Leeson: cover, p. 10; © Reidar Hahn: p. 4, 15, 17, 18; © Bern Pedit: p. 7; © Lynn M. Stone: title page, p. 13, 21; permission of Chicago Historical Society: p. 8; permission of Connecticut Historical Society: p. 12

Library of Congress Cataloging-in-Publication Data

Conlon, Laura, 1959-
Fire / by Laura Conlon.
p. cm. — (Discovery library of disasters)
Includes index.
Summary: Brief text examines the disastrous effects of uncontrollable fires, methods of fire extinction and prevention, and ways to protect people during a fire.
ISBN 0-86593-246-8
1. Fires—Juvenile literature. 2. Fire prevention— Juvenile literature. [1. Fires. 2. Fire prevention.]
I. Title II. Series.
TH9148.C65 1993
363.37—dc20 92-43123
CIP
AC

Printed in the USA

TABLE OF CONTENTS

FIRE

Fire is the heat and light made from material that is burning. Fire's heat and light can be very useful. However, uncontrolled fires can cause **disasters,** major losses of lives and property.

Fire depends upon the presence of **oxygen,** an invisible gas in the air. It also depends upon **fuel,** such as wood or paper, and enough heat to cause burning.

An uncontrolled fire rages during the night

THE HISTORY OF FIRE

About 500,000 years ago, people began to use fire for warmth and cooking. They may have taken flames from fires caused by lightning, a meteor or a volcano.

Early humans learned to rub sticks together to produce a spark to light a fire. Later, the Greeks used a magnifying lens, or burning glass, to direct the sun's rays onto dry wood and start a flame.

In more recent times, people used **tinderboxes** for fires. These metal boxes contained flint and steel and small bits of fuel. When struck together, the flint and steel would create sparks to ignite—or light—the fuel. Matches were invented in 1827.

Flaming lava erupts from a Hawaiian volcano

FIRE DISASTERS

Many fires have caused terrible disasters. The Chicago Fire of 1871, the most destructive fire in United States history, may have started when a cow kicked over a lantern.

The great San Francisco Fire in 1906 and the 1923 Tokyo-Yokohama Fire in Japan were caused by earthquakes. The Tokyo-Yokohama Fire killed 142,000 people.

In 1944, a fire killed or burned 358 people as it raced through a circus tent in Hartford, Connecticut.

The corner of State and Madison lies in ruin after the great Chicago Fire of 1871

FOREST FIRES

Forest fires often start when lightning strikes a tree, or when someone leaves a campfire or cigarette burning. Forest fires burn thousands of acres of woodlands each year in North America.

In 1988 fires in and around Yellowstone National Park destroyed over 1.2 million acres of forest. Some other well-known forest fires have blazed near Tillamook, Oregon; Peshtigo, Wisconsin; and Kenai, Alaska.

The "Black Dragon" forest fire in Siberia and Manchuria destroyed 18 million acres of forest lands in 1987.

A bull elk stands in a charred Yellowstone National Park forest after the 1988 fire

Sixty-nine people died on July 6, 1944, when fire raced through a crowded circus tent in Hartford, Connecticut

Life after death: fireweed begins to brighten the forest floor after a burn in Montana

HELPFUL FIRES

Some forests and prairies, or grasslands, need to be burned to remain healthy. When litter—twigs, leaves and other material—piles up for several years, a really hot, destructive fire may occur because it has so much fuel.

However, when done properly, fires set on purpose by forest and land managers are helpful. These **prescribed burns,** as they are called, help prevent a buildup of forest or grassland litter. Prescribed fires burn quickly. They blacken trees but do not kill them.

This prescribed burn of a prairie marsh helps keep the land healthy

FIRE-FIGHTING EQUIPMENT

Modern fire engines can pump 2,000 gallons of water per minute. Water usually comes from **hydrants,** which are pipes attached to underground water sources.

Fire fighters use ladders to reach high fires and to rescue people. They use axes to break through doors.

Smoke can be as deadly as flames, so fire fighters carry oxygen masks to help them breathe in smoky places. They also wear protective helmets, coats, boots and gloves.

Fire fighters are carefully trained to battle many kinds of fires

PUTTING OUT FIRES

A fire will die out if someone removes the fuel, oxygen or heat. If you had a burning candle, for example, you could remove the oxygen by covering the flame with a glass. You could remove the heat by putting water on the flame, or remove fuel by letting the candle melt away.

Fire fighters use water to put out, or extinguish, most fires. Special chemicals are used on oil and electrical fires.

To fight forest fires, bulldozers are sometimes used to clear trees away from the land around the fire. That removes fire fuel.

Fire fighters pour water on a blazing barn

PREVENTING FIRES

People can do several things to prevent fires. Homes should be cleared of rubbish, especially around fireplaces and furnaces. Old, worn electrical cords should be repaired or replaced.

Homes should be equipped with **smoke detectors,** which will sound an alarm if they sense smoke. A **fire extinguisher** should be available to put out small fires. Everyone should have a fire escape plan.

Learning about fire safety helps prevent fire disasters

AURORA

PROTECTING PEOPLE FROM FIRES

In case of fire, shout a warning. If you are in a building, leave quickly. If fire is around you, escape by crawling, because the cooler air near the floor is safer. If you touch a door and it is hot, do not open it. After leaving the building, call 9-1-1 or the fire department.

Always remember STOP, DROP and ROLL. If your clothes catch fire, don't run. Drop to your knees and roll on the ground to smother the flames.

Glossary

disaster (diz AS ter) — an event that causes a great loss of property and/or lives

fire extinguisher (FIRE ex STING wish er) — an instrument filled with water or chemicals used to put out, or extinguish, fires

fuel (FEWL) — a substance that is burned and releases heat

hydrant (HI drant) — a covered pipe that sticks out of the ground and to which hoses are attached for water

oxygen (AHKS uh jen) — a gas found naturally in the air we breathe

prescribed burn (pre SCRIBED BURN) — a fire lit intentionally for a useful purpose, especially in forests or grasslands

smoke detector (SMOKE de TEK ter) — a device that sounds an alarm when it senses smoke

tinderbox (TIN der bahks) — a small box filled with flint, steel and small bits of material for fuel

INDEX